D1823801

**Hertfordshire**
COUNTY COUNCIL
Libraries, Arts
& Information

# A MATERIAL WORLD

# It's COTTON

KAY DAVIES and WENDY OLDFIELD

Wayland

# A MATERIAL WORLD

It's Cotton   It's Plastic
It's Glass   It's Rubber
It's Metal   It's Wood

Editor: Joanna Housley
Designer: Loraine Hayes

First published in 1993 by
Wayland (Publishers) Ltd
61 Western Road, Hove
East Sussex BN3 1JD, England

**British Library Cataloguing in Publication Data**
Davies, Kay
It's Cotton. – (Material World Series)
I. Title   II. Oldfield, Wendy   III. Series
677

ISBN  0 7502 0856 2

Typeset by Kalligraphic Design Ltd, Horley, Surrey
Printed and bound in Belgium by Casterman S.A.

Words that appear in **bold** in the text
are explained in the glossary on page 22.

# IT'S COTTON

Cotton comes from plants that are grown in many parts of the world. After it has been picked it goes through several changes to become cloth. A lot of your clothes are probably made from cotton. It can be found in different types of fabric, including towels, lace and canvas. The remains of the cotton seed are useful too. Camera films, animal food and cooking oil are made from the parts that cannot be used in fabric. Look around you and use this book to help you to find things that are made from cotton.

Cotton is cool to wear in summer.

It helps keep our bodies comfortable in hot, sticky weather.

Cotton comes from the seed case of a plant. The short, fluffy **fibres** are twisted and **spun** together to make long, strong threads.

After it has been spun, cotton cloth can be **dyed** in many beautiful colours and patterns.

Cotton cloth can be cut into many shapes.

These are joined together on a sewing machine to make our clothes.

7

When our skin is wet the soft loops on a cotton towel easily soak up the water. We soon feel dry and warm.

The rider's denim jeans rub against the hard saddle.

Denim is a strongly **woven** cotton material that lasts a long time.

These Greek costumes are very beautiful.

It takes many months to hand-**embroider** the patterns on to the cotton cloth.

A **lace** shawl can turn ordinary clothes into very special clothes.

Single threads of cotton are woven to make see-through fabric like a spider's web.

These huge beach umbrellas protect us from the sun instead of the rain.

The thin cloth on an umbrella is woven very finely and made waterproof so that rain just rolls off its stretched surface.

**Canvas** tents are made from cotton that is waterproof and wind-proof.

This **marquee** will provide shelter for the guests at an outdoor party.

The bright cotton flags flutter in the breeze.

Every country has its own flag. Do you know what your country's flag looks like?

The film that is used in a camera is made from part of the cotton seed. It is **processed** to give us photographs, which remind us of happy times.

Twists of cotton fibres make soft, strong string. String has thousands of uses. It can make a pretty decoration, like this knotted head-dress.

The string inside the candle burns slowly as the wax melts.

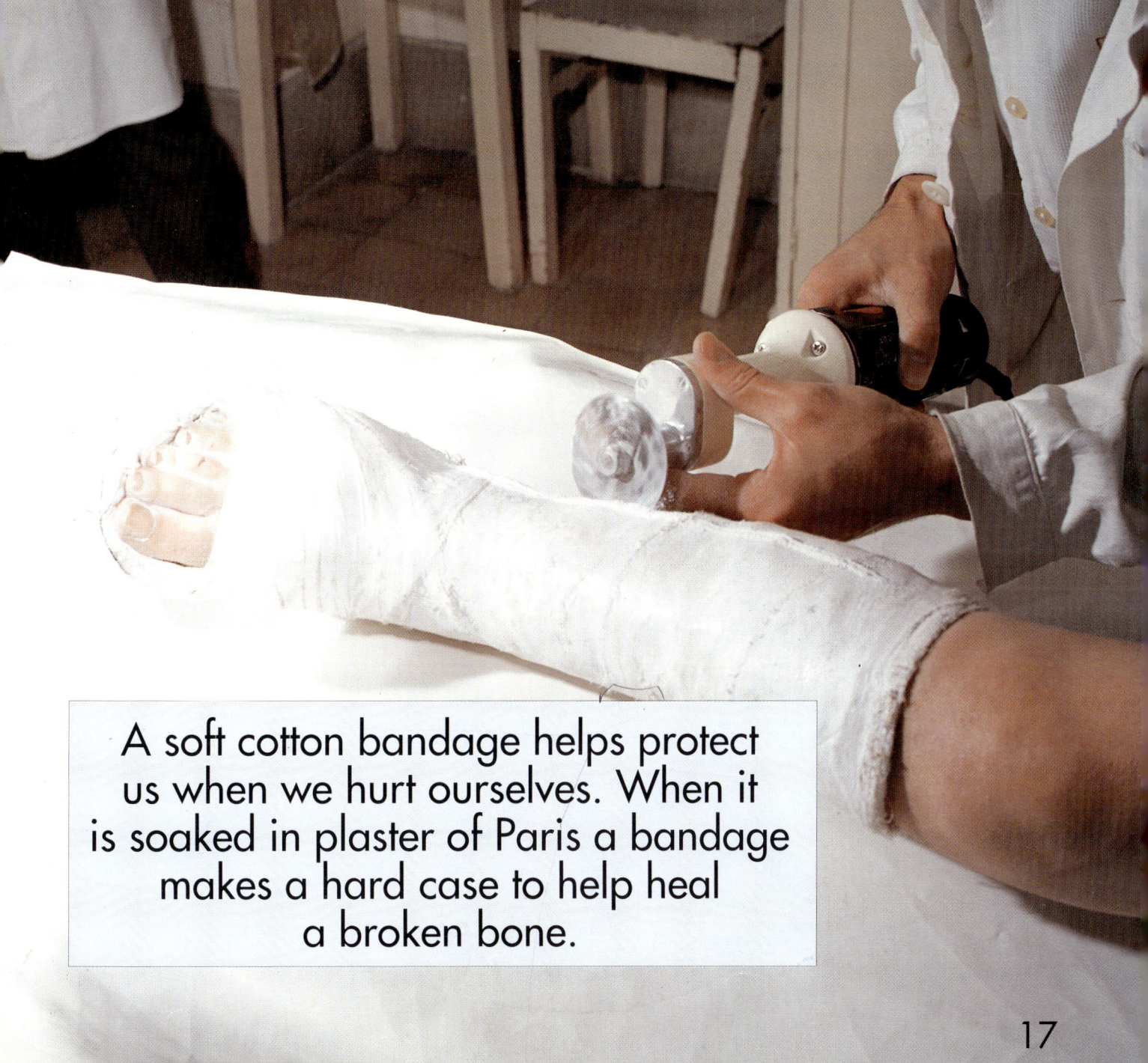

A soft cotton bandage helps protect
us when we hurt ourselves. When it
is soaked in plaster of Paris a bandage
makes a hard case to help heal
a broken bone.

The rag doll
matches
the girl's
dress.

Scraps of
leftover
cloth have
always
been used
to make
toys for
children.

The parts of the cotton plant that cannot be used for fabric are not wasted. The remains of the cotton seed can be used to make sausage skins.

Seed **husks** from cotton
are used in cattle food.
These cows will have
plenty to eat in winter
when the grass is poor.

Old cotton material can be used again.

Paper made from cotton rags is very tough. **Recycled** cotton makes bank notes strong.

# GLOSSARY

**Canvas** A heavy, closely-woven cloth made from cotton.

**Dyed** Material which has been coloured.

**Embroider** To sew a pattern on to cloth.

**Fibres** Single threads of plant or animal material.

**Husks** Outer coverings.

**Lace** A delicate fabric woven in an open pattern. It is used for decoration.

**Marquee** A large tent used for parties held outside.

**Processed** To change a photographic film with chemicals to make pictures appear.

**Recycled** Made into something new from old material.

**Spun** Fibres that have been twisted together to make threads.

**Woven** Made into cloth by crossing over threads.

# BOOKS TO READ

*Clothes* by Linda Howe (Collins Primary Science, 1991)

*Cotton* by Renu Nagrath (A & C Black, 1990)

*Materials* by Kay Davies and Wendy Oldfield (Wayland, 1991)

# TOPIC WEB

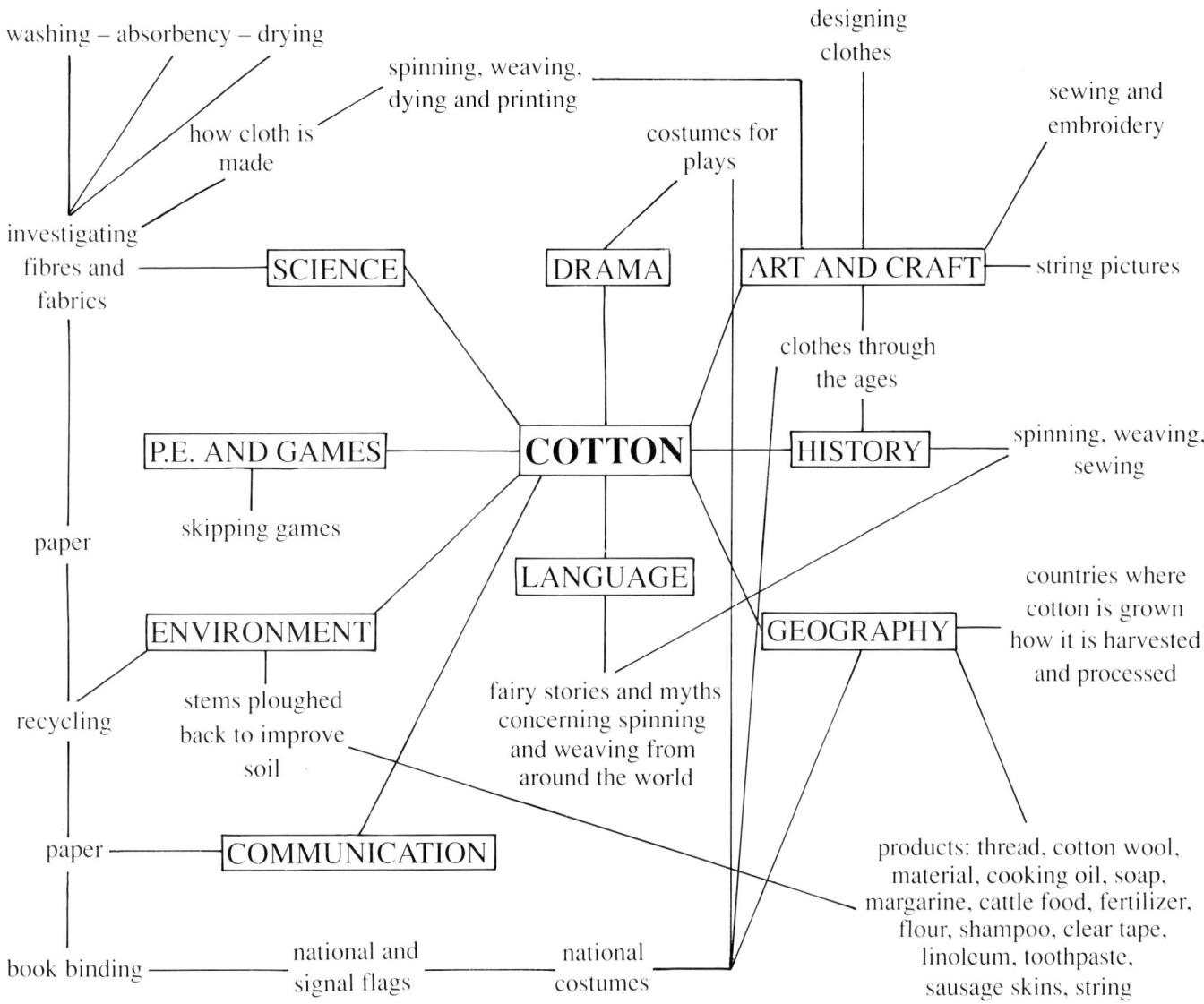

washing – absorbency – drying

how cloth is made

spinning, weaving, dying and printing

designing clothes

sewing and embroidery

costumes for plays

investigating fibres and fabrics

**SCIENCE**

**DRAMA**

**ART AND CRAFT**

string pictures

**P.E. AND GAMES**

**COTTON**

clothes through the ages

**HISTORY**

spinning, weaving, sewing

paper

skipping games

**LANGUAGE**

**ENVIRONMENT**

**GEOGRAPHY**

countries where cotton is grown how it is harvested and processed

recycling

stems ploughed back to improve soil

fairy stories and myths concerning spinning and weaving from around the world

paper

**COMMUNICATION**

products: thread, cotton wool, material, cooking oil, soap, margarine, cattle food, fertilizer, flour, shampoo, clear tape, linoleum, toothpaste, sausage skins, string

book binding

national and signal flags

national costumes

# INDEX

## Picture acknowledgements

The publishers wish to thank the following for supplying the photographs in this book: Cephas Picture Library 18 (Mick Rock); Chapel Studios 13; Eye Ubiquitous 6 (John Hulme), 15 (Paul Seheult), 16 (left, Frank Leather); International Institutue for Cotton 5 (inset); Tony Stone Worldwide *cover* (top), 4 (Nicole Katano, 5 (main pic, Robin Smith), 10, 11 (left, Michelle Garrett), 12 (right, Alena Vikova), 14 (Doug Armand); Topham Picture Source 9, 12 (left); Wayland Picture Library *cover* (right), 7 (both), 8, 19 (A Hasson); ZEFA *cover* (left).